IF FOUND, PLEASE RETURN TO

....

....

....

BEFORE YOU GET STARTED

Thank you for purchasing this notebook. You will find great value in the templates designed to help make your life easier.

Whether you are managing a project for the first time or a seasoned project manager, the goal of this all-in-one notebook is to help keep everything in one place. It is ideal to use one notebook per project.

The notebook is divided into four sections: ***Project Templates, Risks, 2021 Planner,*** *and* ***Meeting Notes***.

1) **Project Templates** contains templates to help capture all necessary information needed to run a project successfully such as defining the problem statement, listing what is out of scope, and capturing high-level milestones to name a few.

2) **Risks** contains templates dedicated to listing all the activities that can compromise the success of your project.

3) **2021 Planner** provides a twelve-month 2021 calendar to help you keep your dates on track.

4) **Meeting Notes** contains templates where you can write your notes, ideas, list whom you met with by date, and actions or decisions made during your meetings.

We welcome your feedback and any additions or changes that come to mind as you use this notebook, please feel free to reach out and send your thoughts to **hello@welbeckadvisors.com.**

ABOUT THE AUTHOR

Bernardo Tirado, PMP is an industrial psychologist and behavioral scientist with Project Management, Six Sigma Blackbelt, Agile Scrum, Body Language, and Executive + Business Coaching Certifications.

Bernardo is a former senior executive for American Express, Morgan Stanley, and JP Morgan Chase and is currently the Managing Partner for Welbeck Advisors (www.welbeckadvisors.com)

Bernardo has taught undergraduate and graduate-level leadership, project management, and industrial psychology courses and is the author of several books including *"Maximizing Project Success Through Human Performance"*, *"Leadership: Helping Others To Succeed"* and contributing author for the *"Gower's Handbook of Project Management"*.

Bernardo shares his monthly thoughts on leadership and technology in his PsychologyToday.com column and his weekly thoughts on career advancement in his podcast and YouTube Channel - Career Insights with Bernardo Tirado.

Feel free to follow Bernardo on Instagram, Twitter, and TikTok:

@ProfessorTirado

ORDER MORE BOOKS USING QR CODE

https://go.professortirado.com/2021Notebook

TABLE OF CONTENTS

I. PROJECT TEMPLATES Pages 7 - 20

- Project Summary [pg. 9]
- Project Scope [pg. 11]
- Measures of Success [pg. 13]
- Project Failure Factors [pg. 15]
- High-Level Milestones [pg. 17]
- Project Team [pg. 19]

II. RISKS Pages 21 - 34

- Managing Risks Templates

III. 2021 PLANNER Pages 35 - 50

- Year-at-a-Glance [pg. 37]
- Monthly Calendar

IV. MEETING NOTES Pages 51 - 98

- Meeting Note Templates
- Action Items
- Notes & Ideas

PROJECT TEMPLATES

PROJECT SUMMARY

Problem Statement

Describe the business reason(s) for initiating the project, specifically stating the business problem.

Estimated Start Date

Estimated End Date

Estimated Budget

Estimated Benefit

Type of Project

Describe the type of project if you manage a range of projects such as changing technology, driving process change, creating a new building, etc.

Project Objective(s)

Describe the business goals and objectives of the project.

NOTES & IDEAS

PROJECT SCOPE

Describe the project scope. The scope defines project limits and identifies the products and/or services delivered by the project. The scope establishes the boundaries of the project and should describe products and/or services that are outside of the project scope.

Project Includes:

Project Excludes:

NOTES & IDEAS

MEASURES OF SUCCESS

Critical Success Factors

Describe the factors or characteristics that are deemed critical to the success of a project, such that, in their absence the project will fail.

Return on Investment

Estimate the orders of magnitude investment for the initiative as well as people hours dedicated to project.

NOTES & IDEAS

PROJECT FAILURE FACTORS

Assumptions

Describe any project assumptions related to business, technology, resources, scope, expectations, or schedules.

Constraints

Describe any project constraints being imposed in areas such as schedule, budget, resources, products to be reused, technology to be employed, products to be acquired, and interfaces to other products. List the project constraints based on the current knowledge today.

NOTES & IDEAS

HIGH LEVEL MILESTONES

List the project's major milestones and deliverables and the target dates for delivery. This list should reflect products and/or services delivered to the end user as well as the delivery of key project management or other project-related work products.

No.	Milestone/Deliverable	Target Date
1		
2		
3		
4		
5		
6		
7		
8		
9		
10		

NOTES & IDEAS

PROJECT TEAM

Summarize roles and responsibilities for the project team and stakeholders identified in the project structure above.

Role	Name
Project Sponsor	
Major Stakeholder(s)	
Project Lead	
Project Manager	
Business Analyst	
Subject Matter Expert	
User-Experience	
Lead Developer	
Reporting and Metrics	
Other	

NOTES & IDEAS

RISKS

MANAGING RISKS

Today's Date

List identified risks, describe potential causes and anticipated impacts, and detail action to mitigate risk along with owner.

Identified Risk(s)

List risk(s) identified

Potential Cause(s)

Anticipated Impact

Mitigation Action

Actions to either either accept, avoid, control, or transfer risk(s).

Owner

Status

✓	ACTION ITEMS	OWNER	DUE
○			
○			
○			
○			
○			
○			
○			
○			
○			
○			
○			
○			
○			
○			
○			
○			
○			
○			
○			
○			
○			
○			
○			
○			
○			
○			
○			
○			
○			
○			
○			
○			

MANAGING RISKS

Today's Date

List identified risks, describe potential causes and anticipated impacts, and detail action to mitigate risk along with owner.

Identified Risk(s)

List risk(s) identified

Potential Cause(s)

Anticipated Impact

Mitigation Action

Actions to either either accept, avoid, control, or transfer risk(s).

Owner

Status

✓	ACTION ITEMS	OWNER	DUE

MANAGING RISKS

Today's Date

List identified risks, describe potential causes and anticipated impacts, and detail action to mitigate risk along with owner.

Identified Risk(s)

List risk(s) identified

Potential Cause(s)

Anticipated Impact

Mitigation Action

Actions to either either accept, avoid, control, or transfer risk(s).

Owner

Status

✓	ACTION ITEMS	OWNER	DUE
○			
○			
○			
○			
○			
○			
○			
○			
○			
○			
○			
○			
○			
○			
○			
○			
○			
○			
○			
○			
○			
○			
○			
○			
○			
○			
○			
○			
○			
○			
○			
○			

MANAGING RISKS

Today's Date

List identified risks, describe potential causes and anticipated impacts, and detail action to mitigate risk along with owner.

Identified Risk(s)

List risk(s) identified

Potential Cause(s)

Anticipated Impact

Mitigation Action

Actions to either either accept, avoid, control, or transfer risk(s).

Owner

Status

✓	ACTION ITEMS	OWNER	DUE
○			
○			
○			
○			
○			
○			
○			
○			
○			
○			
○			
○			
○			
○			
○			
○			
○			
○			
○			
○			
○			
○			
○			
○			
○			
○			
○			
○			
○			
○			
○			
○			

MANAGING RISKS

Today's Date

List identified risks, describe potential causes and anticipated impacts, and detail action to mitigate risk along with owner.

Identified Risk(s)

List risk(s) identified

Potential Cause(s)

Anticipated Impact

Mitigation Action

Actions to either either accept, avoid, control, or transfer risk(s).

Owner

Status

✓	ACTION ITEMS	OWNER	DUE
○			
○			
○			
○			
○			
○			
○			
○			
○			
○			
○			
○			
○			
○			
○			
○			
○			
○			
○			
○			
○			
○			
○			
○			
○			
○			
○			
○			
○			
○			
○			
○			

MANAGING RISKS

Today's Date

List identified risks, describe potential causes and anticipated impacts, and detail action to mitigate risk along with owner.

Identified Risk(s)

List risk(s) identified

Potential Cause(s)

Anticipated Impact

Mitigation Action

Actions to either either accept, avoid, control, or transfer risk(s).

Owner

Status

✓	ACTION ITEMS	OWNER	DUE
○			
○			
○			
○			
○			
○			
○			
○			
○			
○			
○			
○			
○			
○			
○			
○			
○			
○			
○			
○			
○			
○			
○			
○			
○			
○			
○			
○			
○			
○			
○			
○			

2021 PLANNER

2021

JANUARY

S	M	T	W	T	F	S
31					1	2
3	4	5	6	7	8	9
10	11	12	13	14	15	16
17	18	19	20	21	22	23
24	25	26	27	28	29	30

FEBRUARY

S	M	T	W	T	F	S
	1	2	3	4	5	6
7	8	9	10	11	12	13
14	15	16	17	18	19	20
21	22	23	24	25	26	27
28						

MARCH

S	M	T	W	T	F	S
	1	2	3	4	5	6
7	8	9	10	11	12	13
14	15	16	17	18	19	20
21	22	23	24	25	26	27
28	29	30	31			

APRIL

S	M	T	W	T	F	S
				1	2	3
4	5	6	7	8	9	10
11	12	13	14	15	16	17
18	19	20	21	22	23	24
25	26	27	28	29	30	

MAY

S	M	T	W	T	F	S
30	31					1
2	3	4	5	6	7	8
9	10	11	12	13	14	15
16	17	18	19	20	21	22
23	24	25	26	27	28	29

JUNE

S	M	T	W	T	F	S
		1	2	3	4	5
6	7	8	9	10	11	12
13	14	15	16	17	18	19
20	21	22	23	24	25	26
27	28	29	30			

JULY

S	M	T	W	T	F	S
				1	2	3
4	5	6	7	8	9	10
11	12	13	14	15	16	17
18	19	20	21	22	23	24
25	26	27	28	29	30	31

AUGUST

S	M	T	W	T	F	S
1	2	3	4	5	6	7
8	9	10	11	12	13	14
15	16	17	18	19	20	21
22	23	24	25	26	27	28
29	30	31				

SEPTEMBER

S	M	T	W	T	F	S
			1	2	3	4
5	6	7	8	9	10	11
12	13	14	15	16	17	18
19	20	21	22	23	24	25
26	27	28	29	30		

OCTOBER

S	M	T	W	T	F	S
31					1	2
3	4	5	6	7	8	9
10	11	12	13	14	15	16
17	18	19	20	21	22	23
24	25	26	27	28	29	30

NOVEMBER

S	M	T	W	T	F	S
	1	2	3	4	5	6
7	8	9	10	11	12	13
14	15	16	17	18	19	20
21	22	23	24	25	26	27
28	29	30				

DECEMBER

S	M	T	W	T	F	S
			1	2	3	4
5	6	7	8	9	10	11
12	13	14	15	16	17	18
19	20	21	22	23	24	25
26	27	28	29	30	31	

JANUARY 2021

Sun	Mon	Tue	Wed	Thu	Fri	Sat
27	28	29	30	31	1	2
3	4	5	6	7	8	9
10	11	12	13	14	15	16
17	18	19	20	21	22	23
24	25	26	27	28	29	30
31	1	2	3	4	5	6

FEBRUARY 2021

Sun	Mon	Tue	Wed	Thu	Fri	Sat
31	1	2	3	4	5	6
7	8	9	10	11	12	13
14	15	16	17	18	19	20
21	22	23	24	25	26	27
28	1	2	3	4	5	6

MARCH 2021

Sun	Mon	Tue	Wed	Thu	Fri	Sat
28	1	2	3	4	5	6
7	8	9	10	11	12	13
14	15	16	17	18	19	20
21	22	23	24	25	26	27
28	29	30	31	1	2	3

APRIL 2021

Sun	Mon	Tue	Wed	Thu	Fri	Sat
28	29	30	31	1	2	3
4	5	6	7	8	9	10
11	12	13	14	15	16	17
18	19	20	21	22	23	24
25	26	27	28	29	30	1

MAY 2021

Sun	Mon	Tue	Wed	Thu	Fri	Sat
25	26	27	28	29	30	1
2	3	4	5	6	7	8
9	10	11	12	13	14	15
16	17	18	19	20	21	22
23	24	25	26	27	28	29
30	31	1	2	3	4	5

JUNE 2021

Sun	Mon	Tue	Wed	Thu	Fri	Sat
30	31	1	2	3	4	5
6	7	8	9	10	11	12
13	14	15	16	17	18	19
20	21	22	23	24	25	26
27	28	29	30	1	2	3

JULY 2021

Sun	Mon	Tue	Wed	Thu	Fri	Sat
27	28	29	30	1	2	3
4	5	6	7	8	9	10
11	12	13	14	15	16	17
18	19	20	21	22	23	24
25	26	27	28	29	30	31

AUGUST 2021

Sun	Mon	Tue	Wed	Thu	Fri	Sat
1	2	3	4	5	6	7
8	9	10	11	12	13	14
15	16	17	18	19	20	21
22	23	24	25	26	27	28
29	30	31	1	2	3	4

SEPTEMBER 2021

Sun	Mon	Tue	Wed	Thu	Fri	Sat
29	30	31	1	2	3	4
5	6	7	8	9	10	11
12	13	14	15	16	17	18
19	20	21	22	23	24	25
26	27	28	29	30	1	2

OCTOBER 2021

Sun	Mon	Tue	Wed	Thu	Fri	Sat
26	27	28	29	30	1	2
3	4	5	6	7	8	9
10	11	12	13	14	15	16
17	18	19	20	21	22	23
24	25	26	27	28	29	30
31	1	2	3	4	5	6

NOVEMBER 2021

Sun	Mon	Tue	Wed	Thu	Fri	Sat
31	1	2	3	4	5	6
7	8	9	10	11	12	13
14	15	16	17	18	19	20
21	22	23	24	25	26	27
28	29	30	1	2	3	4

DECEMBER 2021

Sun	Mon	Tue	Wed	Thu	Fri	Sat
28	29	30	1	2	3	4
5	6	7	8	9	10	11
12	13	14	15	16	17	18
19	20	21	22	23	24	25
26	27	28	29	30	31	1

JANUARY 2022

Sun	Mon	Tue	Wed	Thu	Fri	Sat
26	27	28	29	30	31	1
2	3	4	5	6	7	8
9	10	11	12	13	14	15
16	17	18	19	20	21	22
23	24	25	26	27	28	29
30	31	1	2	3	4	5

MEETING NOTES

MEETING NOTES

Today's Date

Meeting with

Action(s)

Notes

Decision(s)

✓	ACTION ITEMS	OWNER	DUE
○			
○			
○			
○			
○			
○			
○			
○			
○			
○			
○			
○			
○			
○			
○			
○			
○			
○			
○			
○			
○			
○			
○			
○			
○			
○			
○			
○			
○			
○			
○			

MEETING NOTES

Today's Date

Meeting with

Action(s)

Notes

Decision(s)

NOTES & IDEAS

MEETING NOTES

Today's Date

Meeting with

Action(s)

Notes

Decision(s)

✓	ACTION ITEMS	OWNER	DUE
○			
○			
○			
○			
○			
○			
○			
○			
○			
○			
○			
○			
○			
○			
○			
○			
○			
○			
○			
○			
○			
○			
○			
○			
○			
○			
○			
○			
○			
○			
○			
○			

MEETING NOTES

Today's Date

Meeting with

Action(s)

Notes

Decision(s)

NOTES & IDEAS

MEETING NOTES

Today's Date

Meeting with

Action(s)

Notes

Decision(s)

✓	ACTION ITEMS	OWNER	DUE

MEETING NOTES

Today's Date

Meeting with

Action(s)

Notes

Decision(s)

NOTES & IDEAS

MEETING NOTES

Today's Date

Meeting with

Action(s)

Notes

Decision(s)

✓	ACTION ITEMS	OWNER	DUE

MEETING NOTES

Today's Date

Meeting with

Action(s)

Notes

Decision(s)

NOTES & IDEAS

MEETING NOTES

Today's Date

Meeting with

Action(s)

Notes

Decision(s)

✓	ACTION ITEMS	OWNER	DUE
○			
○			
○			
○			
○			
○			
○			
○			
○			
○			
○			
○			
○			
○			
○			
○			
○			
○			
○			
○			
○			
○			
○			
○			
○			
○			
○			
○			
○			
○			
○			

MEETING NOTES

Today's Date

Meeting with

Action(s)

Notes

Decision(s)

NOTES & IDEAS

MEETING NOTES

Today's Date

Meeting with

Action(s)

Notes

Decision(s)

✓	ACTION ITEMS	OWNER	DUE
○			
○			
○			
○			
○			
○			
○			
○			
○			
○			
○			
○			
○			
○			
○			
○			
○			
○			
○			
○			
○			
○			
○			
○			
○			
○			
○			
○			
○			
○			
○			

MEETING NOTES

Today's Date

Meeting with

Action(s)

Notes

Decision(s)

NOTES & IDEAS

MEETING NOTES

Today's Date

Meeting with

Action(s)

Notes

Decision(s)

✓	ACTION ITEMS	OWNER	DUE
○			
○			
○			
○			
○			
○			
○			
○			
○			
○			
○			
○			
○			
○			
○			
○			
○			
○			
○			
○			
○			
○			
○			
○			
○			
○			
○			
○			
○			
○			
○			

MEETING NOTES

Today's Date

Meeting with

Action(s)

Notes

Decision(s)

NOTES & IDEAS

MEETING NOTES

Today's Date

Meeting with

Action(s)

Notes

Decision(s)

✓	ACTION ITEMS	OWNER	DUE
○			
○			
○			
○			
○			
○			
○			
○			
○			
○			
○			
○			
○			
○			
○			
○			
○			
○			
○			
○			
○			
○			
○			
○			
○			
○			
○			
○			
○			
○			
○			

MEETING NOTES

Today's Date

Meeting with

Action(s)

Notes

Decision(s)

NOTES & IDEAS

MEETING NOTES

Today's Date

Meeting with

Action(s)

Notes

Decision(s)

✓	ACTION ITEMS	OWNER	DUE
○			
○			
○			
○			
○			
○			
○			
○			
○			
○			
○			
○			
○			
○			
○			
○			
○			
○			
○			
○			
○			
○			
○			
○			
○			
○			
○			
○			
○			
○			
○			

MEETING NOTES

Today's Date

Meeting with

Action(s)

Notes

Decision(s)

NOTES & IDEAS

MEETING NOTES

Today's Date

Meeting with

Action(s)

Notes

Decision(s)

✓	ACTION ITEMS	OWNER	DUE
○			
○			
○			
○			
○			
○			
○			
○			
○			
○			
○			
○			
○			
○			
○			
○			
○			
○			
○			
○			
○			
○			
○			
○			
○			
○			
○			
○			
○			
○			
○			

MEETING NOTES

Today's Date

Meeting with

Action(s)

Notes

Decision(s)

NOTES & IDEAS

MEETING NOTES

Today's Date

Meeting with

Action(s)

Notes

Decision(s)

✓	ACTION ITEMS	OWNER	DUE

MEETING NOTES

Today's Date

Meeting with

Action(s)

Notes

Decision(s)

NOTES & IDEAS

MEETING NOTES

Today's Date

Meeting with

Action(s)

Notes

Decision(s)

✓	ACTION ITEMS	OWNER	DUE
○			
○			
○			
○			
○			
○			
○			
○			
○			
○			
○			
○			
○			
○			
○			
○			
○			
○			
○			
○			
○			
○			
○			
○			
○			
○			
○			
○			
○			
○			
○			

Feel free to follow Bernardo on Instagram, Twitter, and TikTok:

@ProfessorTirado

ORDER MORE BOOKS USING QR CODE

https://go.professortirado.com/2021Notebook

www.ingramcontent.com/pod-product-compliance
Lightning Source LLC
Chambersburg PA
CBHW070614220526
45467CB00003B/1428